Electromagnetism *in* Stars Light

Returning to Earth Again Interacting

Janett Lee Wawrzyniak

Copyright © 2011
by Janett Lee Wawrzyniak

First Edition – June 2011
ISBN
978-1-77067-662-6 (Paperback)
978-1-77067-663-3 (eBook)

All rights reserved.
No part of this publication may be reproduced in any form, or by any means, electronic or mechanical, including photocopying, recording, or any information browsing, storage, or retrieval system, without permission in writing from the publisher.

Locations are global known general areas

This book brings the quiet sparkling background of outer space with pulses always traveling to a real home called earth. Space is frequented by electromagnetic life giving wavelengths and it is never totally empty. Presented in continued instruction method information is easier to read. Electromagnetic wave pulses interactions in travel patterns arrive at earth again. Examples through outer space from stars to sun, earth and moon are the paths of interactions. Sights, sounds, reverberation from the observed wave results are seen by all. To then become known expectations, unchanging always there wavelengths support needs and through applied sensors increase quality of life for all.

Published by:

Suite 300 – 852 Fort Street
Victoria, BC, Canada V8W 1H8

www .friesenpress .com

Distributed to the trade by The Ingram Book Company

Contents

Introduction............................... 5

CHAPTER 1................................ 7

    Orion Nebula Lights
    On Orion and Constellation Eridanus

CHAPTER 2............................... 9

    Stars Wavelengths
    On Light Emitting Hydrogen

CHAPTER 3............................. 17

    Solar Energy Converting to Wavelengths
    On Gravitation of Wavelengths

CHAPTER 4............................. 21

    Earths Wavelength System Conversions
    On Volcanic Interactions

CHAPTER 5............................. 25

    Quantum and Strong Nuclear Force
    On Pulse Holding Pattern Light Travel

CHAPTER 6............................. 27

    Electromagnetic Wavelength
    On Wavelength Size

CHAPTER 7............................. 29

    Young Stars
    On Traveling Electromagnetic Radiation

CHAPTER 8............................. 31

Magnetic Solar Flares
On Traveling Electromagnetic Radiation

CHAPTER 9 33

Earths Magnetic Field
On Electromagnetic Sensing Levels

CHAPTER 10 35

Earths Volcanoes
On Volcanic Gasses Interaction

CHAPTER 11 37

Tectonic Plates and Volcanic Activity
On Convergent and Divergent Plates

CHAPTER 12 41

Cyclone, Typhoon and Hurricane Areas
On Storms Formations in
Dense Gravity Field Areas

CHAPTER 13 43

Moon Gravity and Magnetic Fields
On Interaction

CHAPTER 14 45

Quantum Theory, Strong Force,
Weak Force
On Holding Patterns Traveling in Paths

CHAPTER 15 47

Sensors
On Establishing Boundaries
for Specific Use

Additional Information 49

Index 51

Introduction

Within the wealth of information available this book presents electromagnetic wavelengths, their interactions and travel paths. Included are wavelength processes between stars, sun, moon and earth with natural color drawings their shapes indicating direction. How working with quantum the strong nuclear force, larger sight observed areas and their interactions in travel paths with earth functions. Written in continuous instruction method this presentation is easier to understand, including information that can be seen by everyone. The appreciation for traveling wavelengths can be an adventure or discovery. Wavelengths at times flash sparkling pulsing outward into every color and other wavelengths having sound or P waves. Wavelengths are always maintaining what they are. A means of discovery are sensors advancing technologies in wavelengths with some examples given.

CHAPTER 2

Stars Wavelengths
On Light Emitting Hydrogen

Pulsars the neutron star emit a bright light electromagnetic radiation beam with radio waves. The quasar some emitting radio wavelengths, from the quasar electron magnetic center the brightest with intense emissions can be powered by black hole hot regions. The quasar and black hole each have magnetic fields with similar structure and output. A quasar has 1,000 times more power than a galaxy. The compass aligns with a magnetic field. A neutron star is thought to have formed due to a near super nova explosion not to become a black hole but to have the same compact mass. Common hydrogen exists in molecules where young stars or proton stars form, with excited combining of the nebulas electrons gas atoms also then ultra violet high energy light is emitted, a shorter not visible wavelength. Hydrogen does emit light with a wavelength at a higher principal quantum.

CHAPTER 3

Solar Energy Converting to Wavelengths
On Gravitation of Wavelengths

Closer to observe from earth is the sun with its outer solar nebula of volatile gasses converting to wavelengths some in a holding pattern of gravitation. Pulsing quakes in the electromagnetic wavelength system are seen on the sun. Examples of solar energy wave types are acoustic, radio waves, microwaves, x rays, gamma, gravity and sound P mode shapes.

CHAPTER 4

Earths Wavelength System Conversions
On Volcanic Interactions

Observing earth, its wavelength system provides interactions some comparable with other systems throughout the universe. Emissions from volcanoes include hydrogen chloride HCL, hydrogen fluoride HF, hydrogen sulphide H2S, helium He, sulfur SO2, carbon dioxide CO2, carbon monoxide H2, and steam H2O. Pressure in expanding gas explodes the walls of gas bubbles to seen shards or various pumice shapes. Ash is electrically conductive and when wet even more conductive. An example of rapid gas dissolved expansion at Mt. Lassen is in pumice bubble shapes. With hydrogen volcanic processing at times resulting in light can indicate an example of hydrogen with the spectrum of light produced from it in star formation areas.

Mt Lassen Volcano

Mammoth Lakes Volcano

CHAPTER 5

Quantum and Strong Nuclear Force On Pulse Holding Pattern Light Travel

A pulse or excited as hydrogen in a timed transition, at the level of concentration becoming excited energy again the electromagnetic conducting system continually maintains its moment by moment holding pattern. The light speed travel holding pattern of gravitation functioning with a whole universe not regulated to one seen component space, stars, sun, moon, earth as each area is sustaining the whole in connected wavelength gravitation traveling conversions.

Radio 10^4
Size of skyscrapers

Radar 1
Size of humans

Infrared 10^{-2}
Size of coin

Visible 10^{-4}
Size of pin tack

Ultra Violet 10^{-6}
Size of bacteria

X Rays 10^{-10}
Size of atom

Gamma Rays 10^{-14}
Size of atomic nucleus

Electromagnetic Light Spectrum

CHAPTER 6

Electromagnetic Wavelength
On Wavelength Size

Electromagnetic wavelength is a measure used to classify the entire electromagnetic spectrum. Between long wavelength radio transmissions and short gamma rays, increasing in wavelength are microwaves, infrared radiation, visible light, ultra violet light and x rays. At the long wave end of the spectrum are radio transmissions with measured electromagnetic wavelengths that can be the size of buildings. Gamma rays at the opposite end of the spectrum are smaller than an atoms nucleus.

CHAPTER 7

Young Stars
On Traveling Electromagnetic Radiation

Young stars forming in the Orion Nebula with intensive heat cause the region to become more infrared. The Orion Nebula is a part of a larger nebula called the Orion Molecular Cloud Complex. The green hue of Orion is caused by a low probability electron transition in double ionized oxygen. Electromagnetic radiation always travels at the speed of light in the empty space vacuum, and is governed in a vacuum by Maxwell's equations. Black holes or gravity holes, neutron stars are two sources of gravitational radiation waves transporting their energy. Dark matter as in black holes, being older has not been verified.

CHAPTER 8

Magnetic Solar Flares
On Traveling Electromagnetic Radiation

A solar flare is an intense sudden rapid brightness of built up magnetic energy released from the sun emitted across the entire electromagnetic spectrum. A large flare is millions of times greater in energy than the energy released from a volcanic explosion. A flare is a tenth less than the total energy emitted every second by the sun. It takes about seven minutes for electromagnetic radiation to reach earth from the sun. This radiation sustains all life on earth as warm energy.

The solar flare becomes a magnetic storm when the earth's magnetic field is disturbed by coronal mass ejections. The shock wave of solar wind reaching the earth's ionosphere begins the magnetic storm from twenty to thirty eight hours after the solar flare. The magnetic storm can last any number of hours or days. The magnetic storm can disrupt electrical systems. Auroras extending into other areas and intense are indicators a magnetic storm is in progress. When the earth's magnetic and electric process is severely disturbed, by solar storm particles entry into the earths ionosphere and magnetosphere, electric conduction systems become paths for excessive currents created resulting in power failures and cost for equipment replacements. Electromagnetism still continuing, some dates for solar eclipse are given. 2014 Apr 29, an annular eclipse in South Indian and Antarctica areas. 2014 Oct 23 a partial eclipse in the North Pacific and America areas. 2015 Mar 20 a total eclipse in the North Atlantic Faero Islands areas. 2015 Sept 13 partial eclipse in South Africa and Antarctica areas.

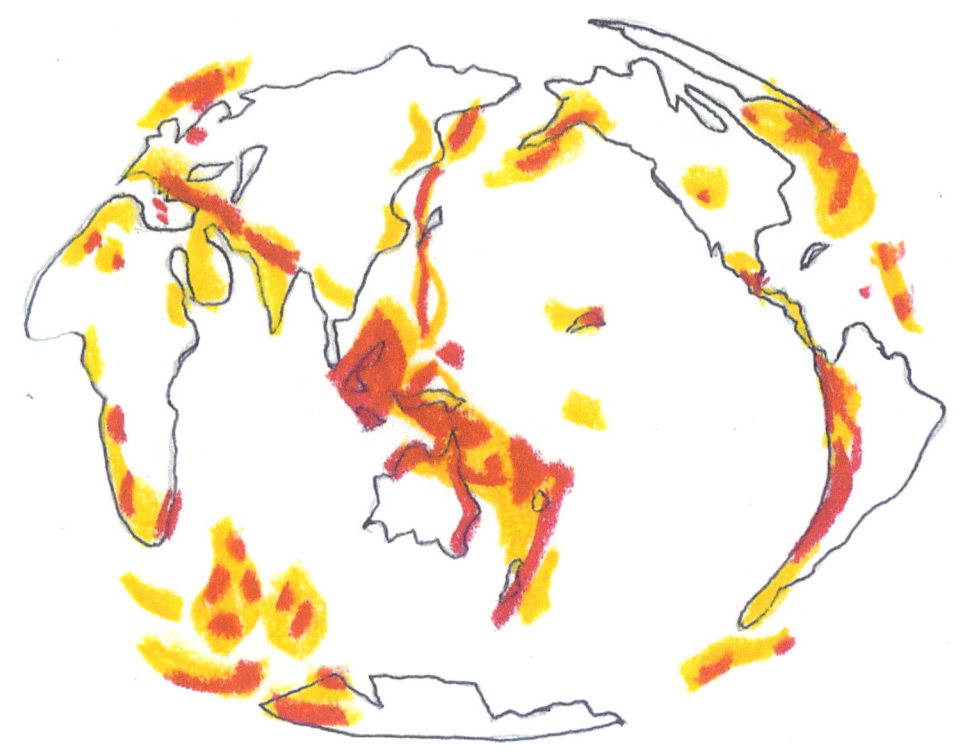

Earths Gravity Field

CHAPTER 9

Earths Magnetic Field
On Electromagnetic Sensing Levels

Earths electromagnetic field has directional effects on small animals having nerve cells in their brains able to process electromagnetic fields to navigation. When magnetic field direction changes certain moles relocate their living area location in coordination with the magnetic field. Turtle's migration routes are with the magnetic field a known direction for navigation. These two examples indicate a behavior pattern. Infrared images of earth indicate areas of more or less intense electromagnetic areas. These areas are located in known earthquake regions.

Many animals are known to leave areas as electromagnetism increases and just before an earthquake. This includes bees leaving their nests, pets becoming more active for no apparent reason, fish moving violently. This increased electromagnetism causes sky fireball lights, migraines, sleep depravation and water changes. One example of water changes with electromagnetic fields is diagnosed cancer patients in the vicinity. Strong magnetic fields cause hallucinations and a sense of feeling exchange, as in warmth from solar energy. Electromagnetic fields can destroy normal functioning DNA and without some supportive chlorella added to diet then continued damaged DNA replication occurs. With normal balance overridden by increased electromagnetic interference brain functions become out of sequence from actual safe occurring events. Recalling passing conversation or an enactment of events not chosen daily the individual, could not be life sustaining. Stress fluxuating at the expense of the individual or individuals in the affected vicinity causes death. Mutated DNA in replication would not respond to normal levels of electromagnetic fields. Additionally there are any number of diseases the altered individual would become susceptible to, which is not considered an exchange rate or life sustaining. Within the known reasonable natural processes or our home called earth, communication by observations of all should indicate areas of safety. When normal daily life is overridden by electromagnetic fields the communicated observation would be to locate in areas of known safety. On our home called earth the electromagnetic increases can be felt in the atmosphere during storms and felt by equipped sensing animals. The electromagnetic spectrum contains P waves or sound waves. P waves from solar activity are also seismic, as in volcanic activity and felt by sensing or equipped animals. If the electromagnetic levels increase the results are sensing fish jumping out of the water, sensing animals evacuating an affected area, a life saving action. Ultra sound a higher frequency is heard also by sensing equipped animals and dolphins. Ultra sound is used as a medical diagnostic tool to be a non hazard mutagenic ionizing radiation. Ultra sound causes inflammatory responses heating soft tissue. Ultra sounds pressure wave causes microscopic bubbles in living tissues distorting cell membranes, ion fluxes and intracellular activity. Ultra sound causes pockets of gas in body fluids or tissues with expansion and ultra sound alters brain function in the electromagnetic spectrum, which does destroy DNA. Ultra sound is used to clean metal by exploding debris to particles releasing them from the intended project.

CHAPTER 10

Earths Volcanoes
On Volcanic Gasses Interaction

Earths volcanoes can cause any number of infections in their vicinity, affecting ear, eye, nose, throat, respiratory and cardiac systems. In the volcanic areas sulfur dioxide destroys DNA and protein in all body cells. Seen in heart tissue, cerebral cortex of brain, lung, liver and kidney. World wide carbon dioxide CO_2 has been reported being concentrated in some volcanic areas, Mammoth Mountain in California is one example. Low lying areas are known collection points for CO_2. Displacement of lake water when a large amount of CO_2 is present can shift the toxic CO_2 gas to populated areas causing asphyxiation. Volcanic hydrothermal systems also indicate CO_2 presence. Hydrogen sulphide H_2S accumulations from volcanic and geothermal sources has resulted in asphyxiation. Malfunctioning geothermal heating systems have been responsible for H_2S asphyxiation also. Hydrogen Sulphide affects nervous, respiratory systems and cardiovascular systems. Sulfur SO_2 also present in active volcanic areas has been responsible in respiratory failure known with existing disease but not childhood asthma. Combinations of gas levels within volcanic activity through gas expansion pressure explodes gas bubble walls into seen pumice shapes.

Blocky pumice shapes are found in Yellowstone when water and lava mix in preheat magnetic, the explosive interaction. Other pumice shapes are bubble-wall and bubble-wall junction. Sensors can be helpful to maintain in place safety. An example is Radon which can be detected and dealt with.

CHAPTER 11

Tectonic Plates and Volcanic Activity On Convergent and Divergent Plates

Convergent or driven together and divergent driven apart tectonic plates are generally considered a cause of volcanic activity. The Pacific Ring of Fire is one convergent plate area and most divergent areas are in the lowest levels of the oceans and Mid Atlantic Ridges. Progressing volcanic activity resulting in carbon dioxide conversion to oxygen by plants is one system example.

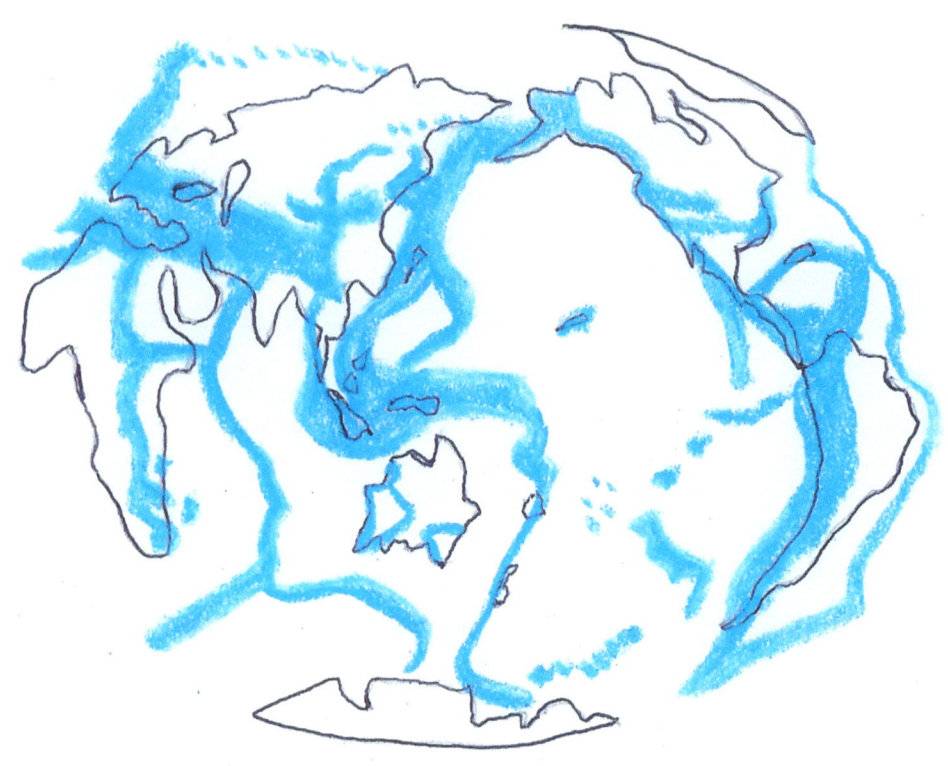

General Earthquake Epicenters

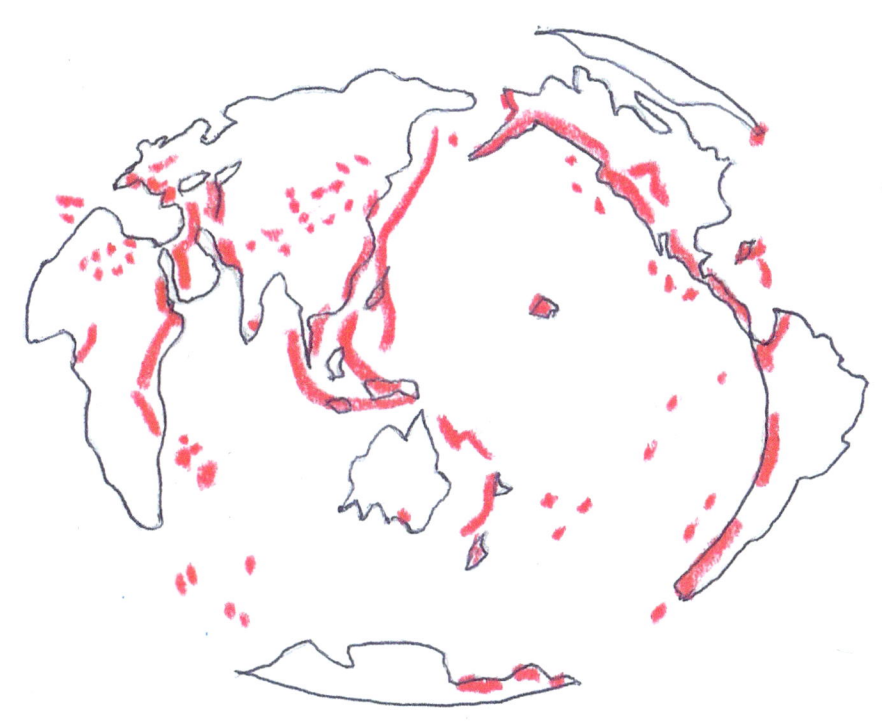

General Volcanic Areas

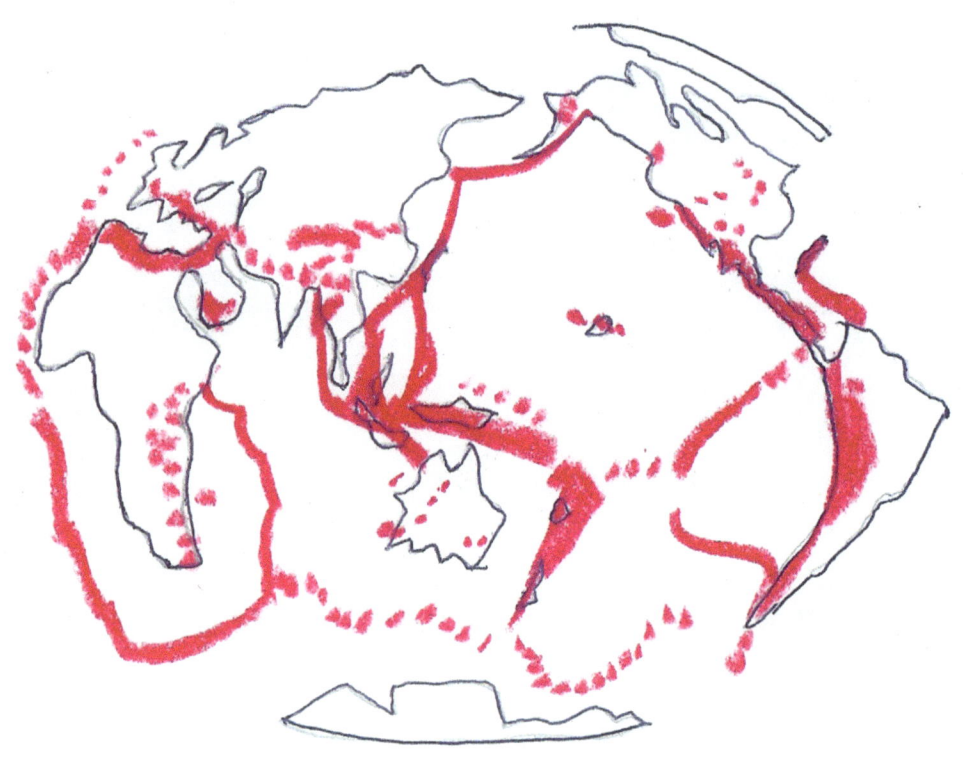

General Earthquake Areas

CHAPTER 12

Cyclone, Typhoon and Hurricane Areas On Storms Formations in Dense Gravity Field Areas

Most Cyclone formations occur in the North Indian Ocean generally between April and December. Typhoon season in the northwest Pacific Ocean begins approximately in May continuing through to November. The eastern north Pacific hurricanes in the central north Pacific begin in June ends around November. Hurricanes in the central north Pacific begin in June ending in November. In the north and central Pacific hurricanes may also be referred to as tropical cyclones. Moving off of the North West African coast are Atlantic atmospheric disturbances called tropical waves. These may form tropical storms and hurricanes when Cyclone, Typhoon and Hurricane areas Intensified by warm temperatures of sea surface and atmosphere. Super size Saharan dust storms are a factor in storm development as they move off of the north western African coast. Dust storms convey dry air centrally in the atmosphere with dust overall. Saharan dust storms support vertical wind shear easterly jets which can stop storm development. Dust in the storms suppresses cloud formation and also can stop storm development.

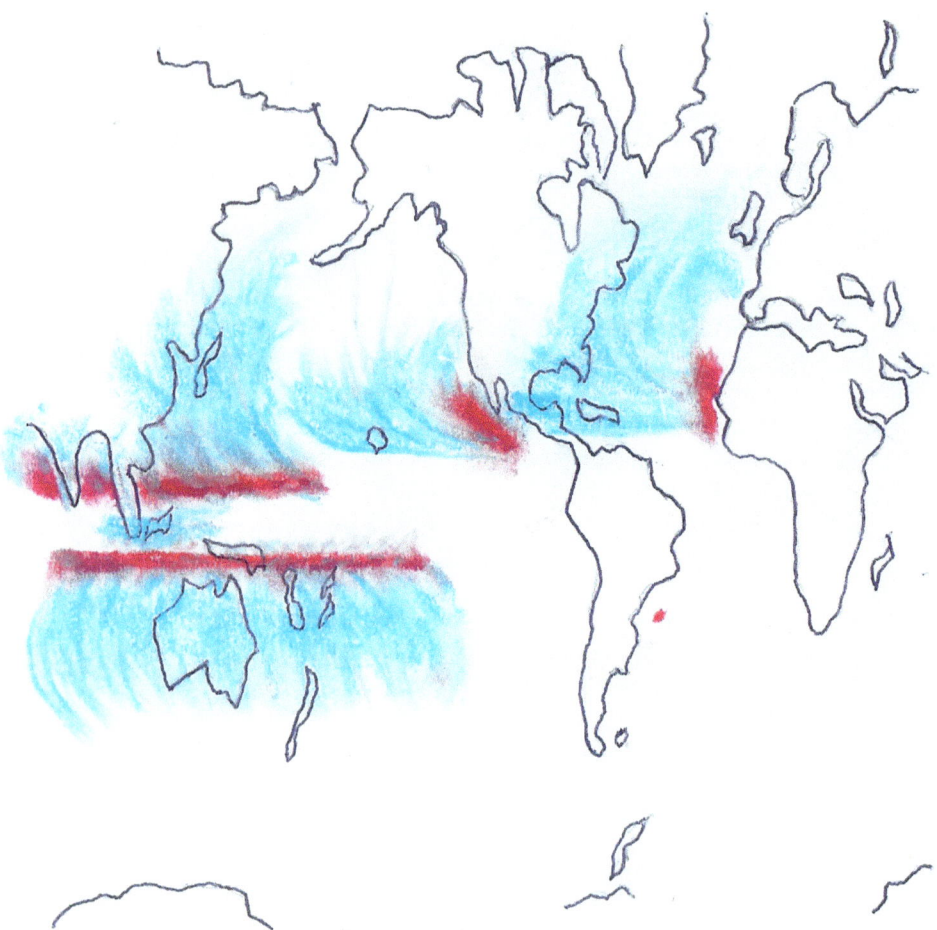

General Storm Origins

CHAPTER 13

Moon Gravity and Magnetic Fields On Interaction

    Measured through the Doppler shift of radio signals from spacecraft, lunar gravity is primarily from some of the larger impact areas, generated at initial contact. The moon magnetic field is less than the earths by approximately one hundredth. Volcanic areas caused by heat producing elements in the vast lunar flats are basaltic lava which flowed into impact craters now having measured magnetization. The moon does not have a measured global dipolar magnetic field generated by a liquid metal core. Lunar atmosphere activity caused by Solar wind ions in contact with Lunar soil resulting in the release of atoms in a near vacuum of variable surface pressure. Water vapor gases have been detected varying with temperature and latitude and may recycle with water ice within moons gravity. Hydrogen in high concentrations is present near the Polar Regions. Lunar atmospheric conditions interacting with geologic activity is so slight that the lunar landscape is well preserved. Other measured activity is solar wind magnetic effects flowing around the lunar magnetosphere.

    The moons surface becomes negatively charged as it passes through the earth's magneto tail. Sun light interacting keeps the lunar build up charges low by repelling some electrons off of the moons surface. Without sun light interaction surface voltages increase to thousands of volts within the lunar night side accumulations. The distance between the earth and moon is increasing. Gravitation between the moon and earths nearest point with the moon reducing earth's energy momentum spin is the cause of increasing distance. This momentum reduction is accelerating the moon into a higher timed orbit resulting in the slowing of Earths spin.

    Interactions of moon and earth rotations are considered to be governing of timed transition stress levels. A measurable observed reaction of stress levels are the moon quakes. The moons size indicates lower measured magnitude of movement lasting longer than an earthquake, the moon and earth in place are with solar effects. When the sun, earth and moon are lined straight eclipses result. Some lunar eclipses are given in 2014 April 15 as total in Pacific, Americas, Australian areas. 2014 Oct 08 as total in Pacific, Americas, Asia and Australia. The next year 2015 April 04 as total in Pacific, Americas, Asia and Australia and 2015 Sept 08 a total in Pacific, Americas, Europe, Africa, and w. Asia.

CHAPTER 14

Quantum Theory, Strong Force, Weak Force
On Holding Patterns Traveling in Paths

The Quantum Theory timed transitions of emission or absorption of energy quantum are connected, in gaseous state elements radiate and absorb light frequencies. Each element absorbs energy when the quantum or amount of energy is reached, emitting or absorbing limited amounts. In the photoelectric effect quantum are called photons, each carrying the exact amount of energy the atom absorbed. The Strong Nuclear Force or strong force holds together subatomic particles of the nucleus, protons carrying a positive charge and neutrons carrying no charge.

The exchange of particles called mesons creates the strong nuclear force. With the meson exchange in place the strong nuclear force holds participating nucleons together. The electromagnetic force can cause repulsion from other proton nucleus. Then to allow meson exchange causing the strong force, two protons nuclei, must be high temperatures for extremely fast movement or under intense pressure allowing close exchange of meson to cause the strong force. Weak Nuclear Force or weak force because its short range field strength is less than electromagnetism.

The four fundamental interaction forces of nature are Strong Nuclear Force, Electromagnetism, Weak Nuclear Force and Gravitation. Gravitational strong fields are described for Planks linearized gravity. Gravitational waves transport energy as gravitational radiation, including neutron stars and black holes as two examples.The string theory connects quantum mechanics and general relativity.

CHAPTER 15

Sensors
On Establishing Boundaries for Specific Use

Radiation damage occurs in sensors, circuits, solar cells as their charges may be as large as incoming radiation ions, electromagnetic storms can also cause damage. Magnetoresistive (MR) sensors determine change in earths magnetic field compassing and navigation and laboratory instrumentation are two uses. Electromagnetic sensors are used in surface vehicles, ships and aircraft. Sensors assist daily with touch screens from computers to cell phones. From smallest and onto larger applications sensors are advancing technology. Sensor improvements directed for quality of life, place many uses in the range of completion. Without sensors various life saving procedures would not be available or in place.

Additional Information

Orion Nebula – an emission nebula, in the north direction having a surrounding luminous cloud, being a hazy object in Orions dagger.

pulsars or neutron star – are pulsating electromagnetic radiation as radio waves, if by short constant interval pulsars are rotating Neutron stars.

quasar – emits high levels of electromagnetic radiation including light emitting more energy than the brightest stars.

black hole – has gravity mass of millions of galaxies drawn in.

super nova – more energetic than Nova, extremely luminous, radiating a lifetime of suns energy.

hydrogen – a highly flammable gaseous element, the lightest of all gasses and most abundant element in the universe.

ultra violet light – energetic radiation emitted by stars ionizing molecules a disruptive effect of Ultra Violet radiation are damages to DNA molecules and can turn green earth areas into barren deserts.

sun solar nebula – energy expelled in thermal pulsations.

HCL – infrared spectrum gaseous at room temperature, most molecules are in the ground vibration state.

HF – found in infrared spectroscopy.

H2S – found in volcanoes and some hot springs and affects the nervous system.

HE – gaseous occurring in atmosphere of sun, stars, earth and spectroscopially dected, is a product of radium emanation.

SO2 – occurs in volcanic gasses and some warm springs, poisonous.

CO_2 – emitted by some volcanoes during sub aerial eruptions, is a product of hot springs and some deep lake water and is poisonous.

H_2 – occurs in molten volcanic rock and difficult to measure.

H_2O – with electrical attraction called hydrogen bonding, constantly moving, gaseous it is steam or vapor.

wavelength – is measured distance between crest or trough of adjacent waves.

electromagnetic gravity – the holding force of elections and protons inside atoms changing electric field generates magnetic field. As in generators, transformers, motors and other uses.

string theory – indicates possible direction toward multi universes and extra dimensions with variations some interacting.

Index

African Coast 43.
animals 35.
auroras 33.
black hole 12, 17, 31, 47, 51.
brain functions 35.
carbon dioxide CO2 25, 37, 41, 52.
carbon monoxide H2 25, 52.
convergent 41.
conversions 27.
converting wavelength 21.
cyclone 43.
dark matter 9, 31.
directional 35.
diseases 35.
divergent 41.
DNA 35, 36.
doppler shift 45.
earth 20, 27, 33, 37.
earth's gravity field 32.
earthquake 35.
earthquake epicenters 38.
electr cally conductive 25, 33.
electromagnetic 27, 33, 35.
electromagnetic gravity 52.
electromagnetic radiation 17, 31.
electromagnetic spectrum 26, 35.
electromagnetic storms 49.
electromagnetic wavelength 21.
electron magnetic center 17.
electron transmission 29.
elements 47.
emissions 25.
emitted 33.
emitted radiation 31.
energy 31.
expanding gas 25.

galaxies 9, 13, 17.
general storm origins 43.
geothermal 37.
gigantic hole 7, 9.
gravitation 27, 45, 47.
gravitation holding pattern 21.
gravitational radiation 31, 47.
gravitational waves 47.
helium He 25, 51.
higher principal quantum 17.
holding pattern 27.
hurricanes 43.
hydrogen 17, 27, 45, 51.
hydrogen chloride HCL 25, 51.
hydrogen fluoride HF 25, 51.
hydrogen sulphide H2S 25, 37, 51.
incoming radiation 9.
infrared 29, 35.
interactions 25.
intracellular activity 36.
light spectrum 25.
lunar atmosphere 45.
lunar eclipses 46.
lunar gravity 45.
lunar magnetosphere 45.
magnetic energy 33.
magnetic fields 17, 33, 35, 45.
magnetic storm 33.
magnetization 45.
magneto tail 45.
Mammoth Mountain 37.
Maxwell equations 31.
meson 47.
Mid Atlantic Ridges 41.
moon 27.
moon magnetic field 45.
moon quakes 45.

Mt. Lassen 22, 25.
mutated DNA 35.
nucleons 47.
neutrons 47.
North Indian Ocean 43.
North West Africa 43.
nuclei 47.
nucleus 47.
Orion Nebula 6, 9, 31, 51.
oxygen 31.
P waves or
sound waves 21, 36.
Pacific Ocean 43.
Pacific Ring of Fire 41.
photoelectric effect 47.
photons 47.
Planks linearized gravity 47.
preheat magnetic 37.
protons 47.
pulsars neutron star
10, 17, 31, 47, 51.
pulsing 21, 27.
quantum theory 47.
quasar 11, 17, 51.
radiation 49.
radio waves 17.
Saharan dust storms 43.
seismic 36.
sensing 36.
sensors 49.
solar 45.
solar activity 36.
solar eclipse 33.
solar energy 35.
solar flare 33.
solar nebula 21, 51.
solar storm 33.
solar wave types 21.
solar wind 33, 45.
space 27.
stars 27.
steam H2O 25, 52.
strong nuclear force 47.
sulfur SO2 25, 37, 51.
sulfur dioxide 37.

sun 18, 19, 27.
sun light 45.
super nova 15, 51.
tectonic plates 41.
timed orbit 45.
timed transition 27, 45, 47.
tropical storm 43.
typhoon 43.
ultra violet 17, 51.
ultra sound 36.
universe 27.
volcanic 33, 36,
37, 39, 41, 45.
volcanoes 22, 23, 27.
voltages 45.
water ice 45.
water vapor gasses 45.
wavelength 25, 52.
wavelength gravitation 27.
weak nuclear force 47.
Yellowstone 37.
young star 14, 17, 31.

www.ingramcontent.com/pod-product-compliance
Lightning Source LLC
Chambersburg PA
CBHW051215220526
45473CB00003B/1041